Answer Sheets of Forestry Courses with A and B Grades in USA

Kulvir Bangarwa

\multicolumn{3}{|c|}{**Table of Contents**}		
Sr. No.	**Topic**	**Page No**
	Preface	**3**
1	**Answer Sheets of Introduction to Forestry Course Examination with B Grade in USA**	**7-21**
2	**Answer Sheets of Tree Improvement Course Examination with A Grade in USA along with General Information and Course Outline**	**23-50**
3	**Answer Sheets of Tree Physiology Course Examination with B Grade in USA**	**51-63**

Preface

Author participated in Indian Council of Agricultural Research (ICAR)-The United States Agency for International Development (USAID) sponsored Training Programme in Professional Forestry Education at Michigan State University, Lansing, USA from 10 September 1987 to 01 September 1988. Author got opportunity to complete 21 credits of Forestry Courses with letter grade and Over All Grade Average of 21 credits was 3.42/4.00. Author was searching his old documents and he was able to search three of Answer sheets of Introduction to Forestry, two of Answer sheets of Tree Improvement along with General Information and Course Outline and two of Answer sheets of Tree Physiology. The marks obtained by author in seven Answer sheets of Introduction to Forestry, Tree Improvement and Tree Physiology ranged from 69 per cent to 100 per cent. Fortunately, Evaluators of Answer sheets have given solution/appropriate suggestion for the wrong Answers. Author has compiled the Answers sheets of three Forestry Courses as "Answer Sheets of Forestry Courses with A and B Grades in USA". Author is thankful to Indian Council of Agricultural Research (ICAR), The United States Agency for International Development (USAID), CCS Haryana Agricultural University, Hisar and Michigan State University, USA.

Dr Kulvir Bangarwa
Retired Professor
CCS Haryana Agricultural University,
Hisar, India

1 Answer Sheets of Introduction to Forestry Course Examination with B Grade in USA

30/30

Quiz #1

Intro to Forestry FOR 202   Name: Kulwir Singh Bangarwa

1. Circle on the map below the location of the six major forest regions in the United States and identify each by name. (12 pts.)

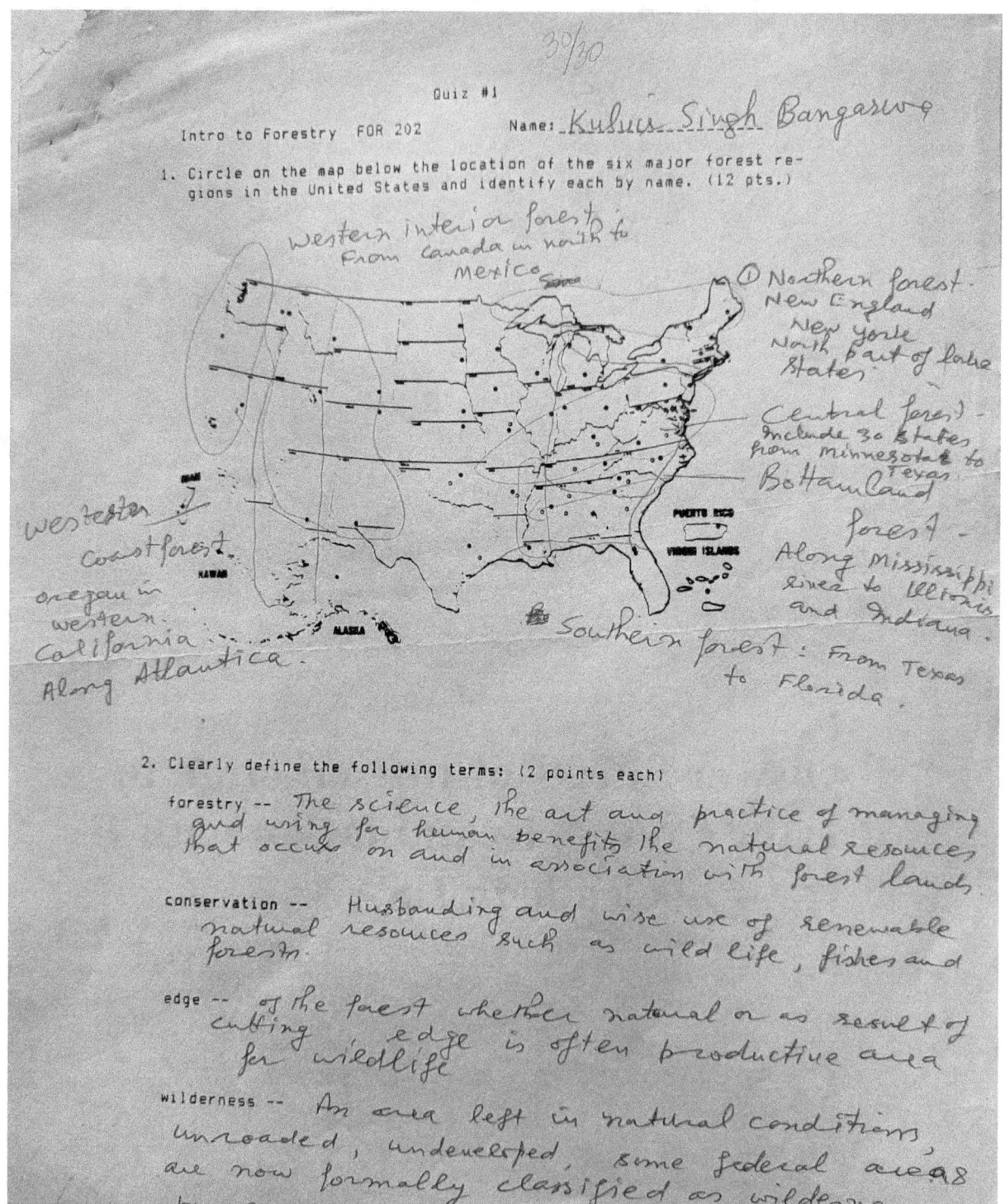

Western interior forest, From Canada in north to Mexico

① Northern forest - New England, New York, North Part of Lake States

Central forest - Include 30 states from Minnesota to Texas

Bottomland forest - Along Mississippi river to Illinois and Indiana.

Western / Coast forest, Oregon in western California Along Atlantica.

Southern forest: From Texas to Florida.

2. Clearly define the following terms: (2 points each)

forestry -- The science, the art and practice of managing and using for human benefits the natural resources that occurs on and in association with forest lands.

conservation -- Husbanding and wise use of renewable natural resources such as wild life, fishes and forests.

edge -- of the forest whether natural or as result of cutting. edge is often productive area for wildlife

wilderness -- An area left in natural conditions, unroaded, undeveloped, some federal areas are now formally classified as wilderness area by congressional act.

3. Define the differences between hardwood, softwood, conifer, evergreen and deciduous trees. (2 points)

Hard wood	Soft wood	Conifer ~~evergreen~~ ~~Deciduous~~
wood produced from deciduous trees. Maple, oak.	wood of conifers. evergreen. fir, pine, cedar.	

4. List three of five guidelines important in maintaining wildlife habitats on forest lands. (6 points)

① ~~clea~~ Keep clearcut of 20 hactares or less.
2. Keep Buffer strips of trees along streams or highways.
3. Management of forest should be prescribed in such a way that burning can be used.
4. Keep the over mature or even diseased

5. What was the significance of the broad arrow mark on trees in colonial times? (2 points)

King broad arrow mark /|\

Best tree having 60 cm diameter 30 cm above ground.
Best tall tree
In 1658 after the act of 1691 kings broad arrow marke was used on the best tree. The tree was reserve for king and allowed to cut by individual. Penalty was castaration.

Introduction to Forestry (For 202)　　　　　　　　　　　　　Fall, 1987

Fall 1987　　　　　　　　　Examination #1
　　　　　　　　　　　　　NAME: K.S. Bangarwa　(1 point)

Directions: Define the following terms in one or two sentences. (2 points each)

1. **farm game:** Early farms with small fields and diverse crops as well as waste and marshy areas are the habitats of farm game animals.
   eg. quail, Fox squirrel, Cotton tail rabbit, Mourning dove

2. **watershed:** The total area contributing drainage to stream.

3. **nonconsumptive-use:** Looking at recreating among natural resources rather than removing for commercial gain or food usually used in connection with wildlife interaction.

4. **multiple-use management:** The administering the forested lands for the purpose of providing outdoor recreation, rangeland, timber and watershed protection and fish and wildlife enhancement for best combination of uses and coordinated management without impairment of the productivity of land with the consideration being given to relative resource value and not the maximum dollar return.

5. **carrying capacity:** The number of animals an area can support adequately with deterioration (degradation) of its site.

6. **cover type:** The species or the mix of tree growing in a geographic area.

7. **hydrologic cycle:** The progress of water from lake or river into rain thence to ground and back to lake again.

Introduction to Forestry (For 202) Fall, 1987

Examination #1

Directions: Answer the questions in one or two sentences.

8. Name two land laws which were part of the first of two important groups of land laws enacted between 1841 and 1871, and explain their importance in relation to speeding settlement in the western territories of the time. (5 points)

Pre emption act 1841 → It provides the head of the family with certain qualification to purchase cheaply 160 acre land for personal settlement, gain and use.

Home stead act 1862

↳ This is broader than Pre emption. It provides free patent after five years of residence and cultivation.

These law speeded up the development of wide spread transportation and settlement. Settlement boundaries extended.

9. The second set of land laws enacted between 1841 and 1871 were passed to serve a different purpose than the first. What was the purpose of this second set of land laws? (2 points)

To maintain the forest or To avoid exploitation of forest.

10. List the five major national forest uses. (5 points)
   1. Recreation
   2. Timber production
   3. Watershed protection
   4. Range management
   5. Wildlife habitate

11. What is the importance of forest litter and humus layer thickness on soil water retention in the forest? (4 points)

More thickness of litter and humus layer more water will holds.

Introduction to Forestry (For 202) Fall, 1987

Examination #1

12. Describe why cattle grazing in a forest woodlot can be a problem? (2 points)

Cattle grazing damage the reproduction by consumption and physical damage to existing trees, which tree seldom recover. ~~Cattle~~ Animal cause holes by hooves causing ~~free~~ entry of tree root diseases.

13. Decribe the grazing sequence for cattle in the west and the reason(s) why the size of herds is regulated by the ~~winter~~ summer range. (5 points)

−3

Minimum in Winter — semi-arid lowlands
Maximum in Summer — high altitudes
Middle in Spring and Fall — mid range of elevations —
Because of forage production
↳ less in summer range ~~therefore~~ and easily ruined w/ overgrazing.

14. List five of seven governing factors which affect the ability of a forest ecosystem to support various populations of wildlife species. (5 points)

Forest type
Tree species composition
Length of tree cutting rotation
Successional stages
Management of forest like
 (a) Burning
 (b) Chemical use.

15. Write the correct landowner (public, nonindustrial private and industrial private forest land ownership) into the correct section of the pie divided below. (3 points)

— Public
— Nonindustrial private
Industrial private

Introduction to Forestry (For 202) Fall, 1987

Examination #1

16. List three important factors related to wildfires which explains why many wildlife species can continue to live in the areas where a burn has recently occurred. (5 points)

−3 At the time of fire wildlife shift and islands of timber return back after burning.
varied intensity of burn – a tree burn
new vegetation next — wild life get space.
size of burn
shape of burn

Directions: Match the most characteristic major tree species listed in the right column with its proper forest region listed in the left column. (2 points each)

Forest Region

15. h Alaskan forest
16. e Western Interior Forest
17. b Central Forest
18. j Northern Forest
19. g West Coast Forest
20. c Southern forest
21. a Tropical forest
22. d Bottomland forest
23. k Boreal forest

−6

Major Species

a. no tree species listed
b. white oak
c. loblolly pine
d. baldcypress
e. ponderosa pine
f. lodgepole pine
g. Douglas-fir
h. Sitka spruce
i. koa
j. eastern white pine
k. white spruce

Directions: Answer these questions with a True or False. (1 point each)

40. False The primary reason that western watersheds are managed is to improve the amount of high quality range available for grazing.

41. False An 80% drop in wildlife populations occurs following a natural wildfire.

−1 42. False In some areas of the west, over 80% of the annual percipitation is snow.

43. False Less than 75% of the U.S. population's recreation time is oreinted toward natural resources.

44. False Wilderness areas, by their very nature, preclude the need for management.

45. False Food quality for wildlife is probably more important than any other factor in habitat management.

Fall 1987   55/80 = 69%   Examination #2
Introduction to Forestry (For 202)

Name __K.S. Bangarwa__

Directions: Define the following terms in one or two complete and legible sentences. (2 pts each)

1. Primary succession: Growing of trees in a bare soils.

2. Secondary succession: Invasion of trees in the soil that is previously vegetated a growing of trees in soil where there is already vegetation.

3. synecology: The study community organism with respect to their environment.

4. forest ecology: Study of interaction occuring among community of forestry may be minute to magnificantly huge.

5. site index: The growth ability of soil is known as soil productivity which is expressed as site index.

6. net growth: → Total growth − Mortality.

Net Growth = 150 + 2 − 14 = 138   Mortality = 14

Net growth will be 138

7. photoperiodism: The period of photosynthesis or Day sunlight. The period in which light is produced. It increases and decreases.

8. principle meridian: North South

    (diagram showing N-S axis labeled "Principle Meridian")

9. DBH: Diameter breast height
   Diameter measured at the breast height
   breast height = 4.5' = 4½ ft.

10. basal area: The area that is actually covered by trees.

    (diagram of square with circles labeled "Trees" and "Basal area"; note: "This area not under basal area")

Directions: Use TRUE or FALSE to indicate the accuracy of the following statements. (1 point each)

11. True Tolerant species tend to exhibit better growth on sites with low site index.

12. True The initiation of fall color is triggered by cool nights in late summer and early fall and bright sunny days.

13. False CAI means Current Annual Index

14. False A face cord encompasses 128 cubic feet of space.

15. True Mycorhizae is a fungus which infects tree roots and helps the tree collect nutrients.

16. False The pholeom carries water and nutrients to the trees branches and leaves.

Directions: Answer the questions in two or three complete and legible sentences.

17. Draw and label the major divisions in a forest soil profile. (5 pts)

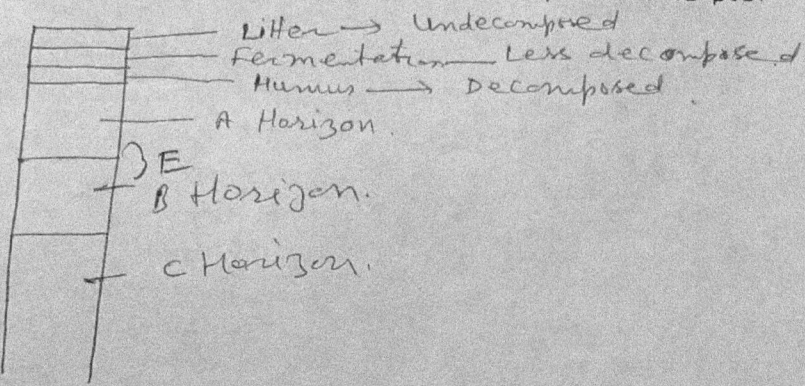
- Litter → Undecomposed
- Fermentation → Less decomposed
- Humus → Decomposed
- A Horizon
- E
- B Horizon
- C Horizon

18. Explain the law of the minimum. (5 points)

The optimum level of nutrients for the proper growth. or The minimum amount of nutrients for a maximum utilization.
Suppose requirement of N is 40 kg N. If we put 50 kg, 10 kg additional will not give economic gain in terms of growth.

19. Describe the movement of cattle in the west in response to the change in seasons and explain why the herds are moved. Also explain why the size of the herd is regulated by the summer range. (5 points)

Herds moves because of shortage of fodder. In summer the return of herd will start because of fodder production

20. List the five major soil forming factors. (5 points)

1. climate
2. Parent material
3. Topography
4. Vegetation
5. Time

Directions: Follow the directions given in each of the following questions.

21. Given the drawing and the dimensions below, how many board feet are ~~there~~ in the sample. Write the volume of wood contained in a board foot. (2 points)

10 Feet, 1 Foot, 6 inches

$= \frac{5}{\cancel{128}}$ bd Ft

60

$10 \times 1 \times \frac{\cancel{6}\,5}{\cancel{12}\,2} = \frac{1}{128}$

Bd Foot volume: $\frac{144 in^3}{5\ cubic ft}$

—2—

22 – 25. Which of the following trees would be counted in and which would be counted out in a variable radius plot established with a 1 inch wide piece of paper held 33 inches from your eye? Circle one each for A., B. and C.? Explain why C is counted in or is counted out. (4 points)

22. A. In **Out**   In
23. B. **In** Out   out
24. C. In Out In Explain why?

It just coincides

25. Label the following diagram. (5 points)
(Note: Also remember to label the heartwood + the sapwood)

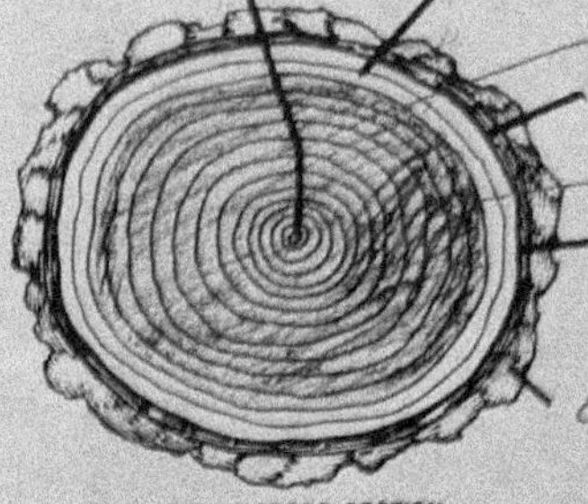

Labels on diagram:
- pith
- xylem cambium
- Sapwood
- cambium
- Heartwood
- Inner bark (phloem)
- outer bark

TRUNK CROSS-SECTION

26. Label the crown classes in the forest illustrated below: (4 points)

#26

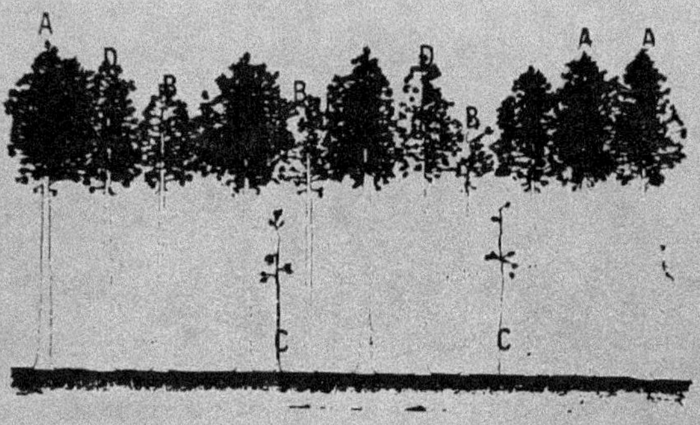

A. Dominant   B. Intermediate

C. Supperessed   D. ~~Supperessed~~ Codominant.

Page 5

Directions: Match the diagram of a leaf with the appropriate term in the column on the left side of the page. (1 point each)

27. __B__ simple
28. __D__ dentated
29. __C__ compound
30. __E__ opposite
31. __A__ alternate

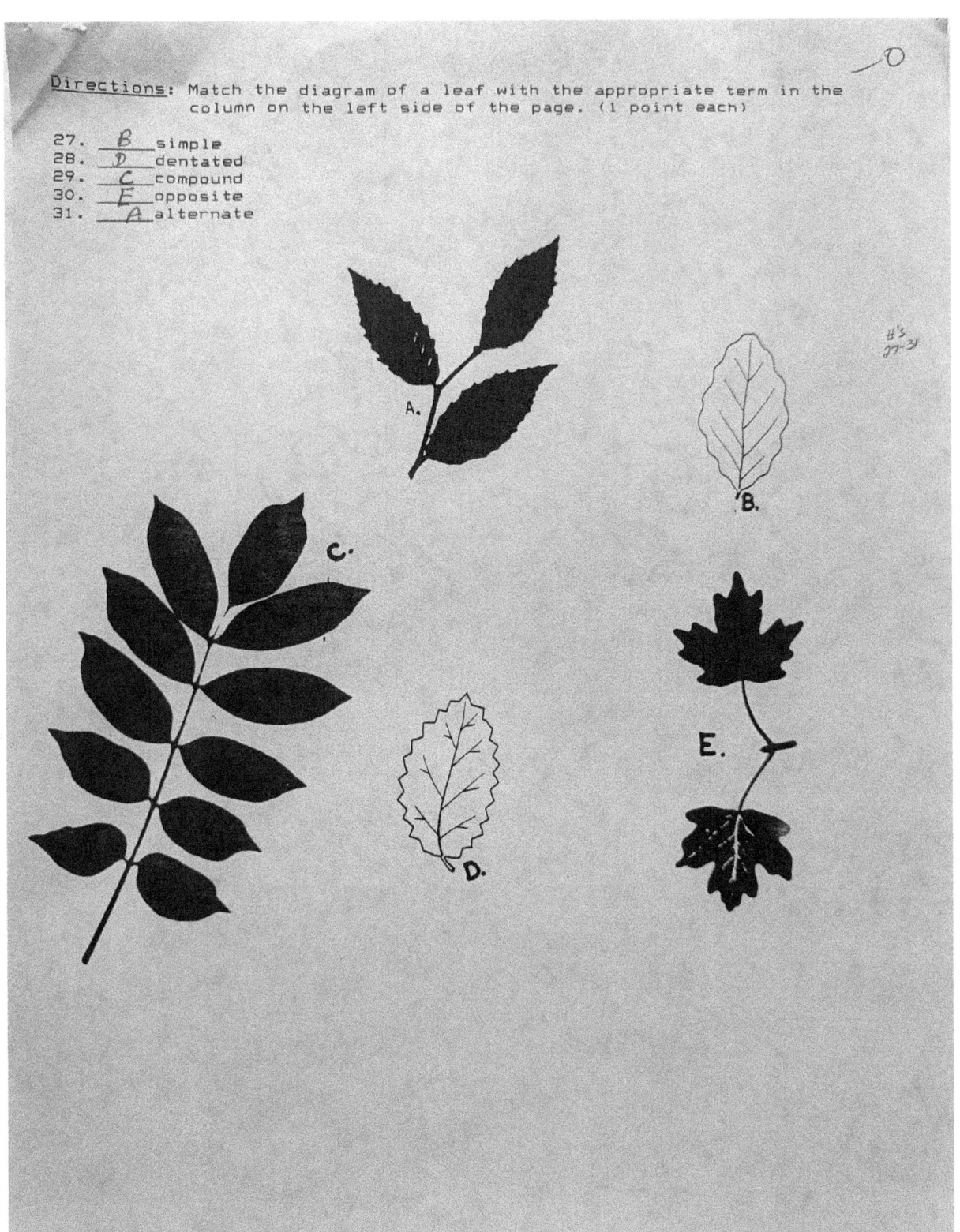

32. Diagram a township and label the various land units which are utilized in the rectangular survey sytem. Size in miles and/or acres should also be included as part of the diagram. (8 points)

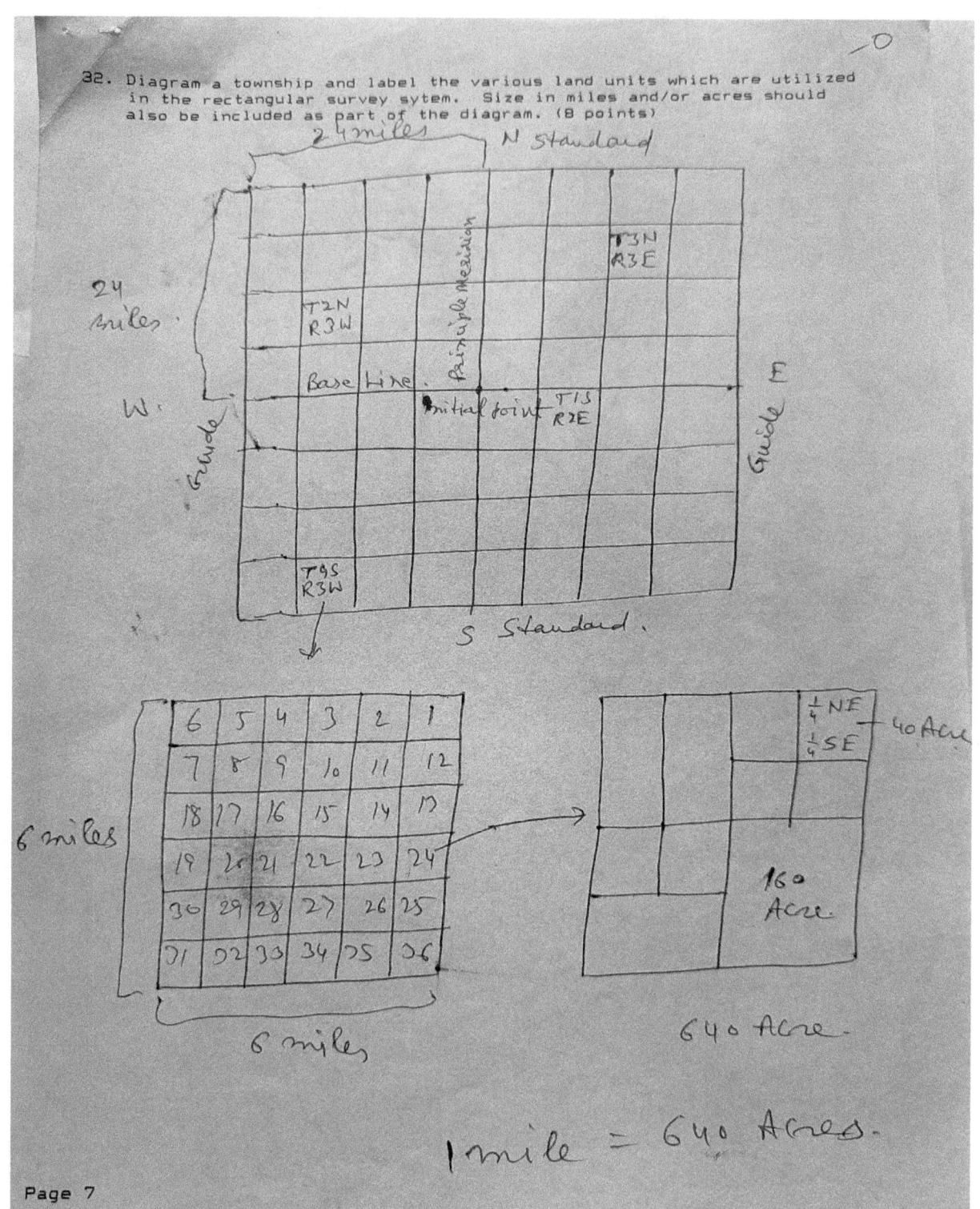

33. Diagram a slope and label the components beginning with the terrace and ending with the summit. (4 points)

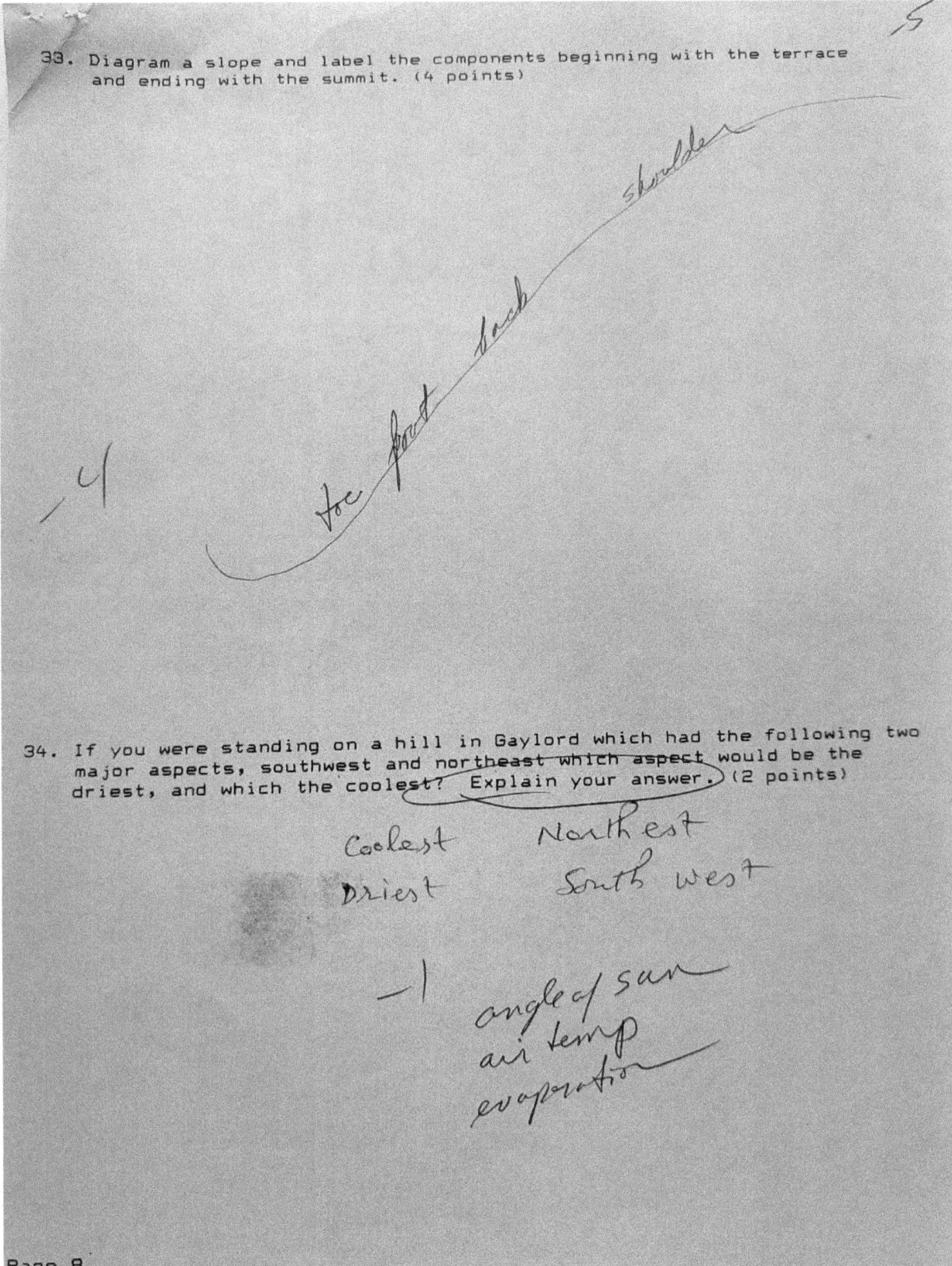

toe, foot, back, shoulder

−.4

34. If you were standing on a hill in Gaylord which had the following two major aspects, southwest and northeast which aspect would be the driest, and which the coolest? Explain your answer. (2 points)

Coolest   Northest
Driest   South west

−1   angle of sun
     air temp
     evaporation

2 Answer Sheets of Tree Improvement Course Examination with A Grade in USA along with General Information and Course Outline

FOR 410 3(2-2)
Tree Improvement
Fall Term 1987

General Information and Course Outline

Instructor: James W. Hanover Office: 119 Natural Resources Bldg.
Office Hours: by appointment   Lab: 102-103 Natural Resources Bldg.
Telephone: 355-0356

Description of Course:

The practical applications of the science of genetics and principles of plant breeding to achieve genetical improvement of tree species.

Objectives of Course:

To develop an understanding of the following topics in the area of tree improvement:

1. Theoretical bases for qualitative and quantitative variation in important traits in trees.

2. The nature and extent of variation in natural populations of tree species.

3. Application of selection methods for the improvement of tree traits.

4. Techniques for maximizing the degree of genetic improvement in a given species.

5. Procedures for accomplishing all phases of a practical tree improvement program.

6. Methods of evaluating and justifying tree improvement efforts for a species.

References:

Text: Zobel, Bruce and John Talbert, 1984. Applied forest tree improvement. John Wiley, N.Y. 505 p.

See list of references attached.

Attendance Requirements: None. However, exams depend substantially on lecture materials.

FOR 410

<u>Grading and course requirements</u>: Your grade for the course will be based upon the following criteria:

2 one hour exams	40%
Final Exam	25%
Term Report	25%
Life Cycle Drawing	5%
Field and Lab Notes	5%
	100%

Grades will be assigned according to the following averaged performance:

91 - 100%	4.0		71 - 75%	2.0
86 - 90	3.5		66 - 70%	1.5
81 - 85	3.0		61 - 65	1.0
76 - 80	2.5		60 - less	0.0

FOREST TREE IMPROVEMENT

Topic Outline

	Week	Text Readings
1. Introduction to tree improvement	1	ch. 1
2. Mendelian genetics	2	ch. 2 (39-58)

3. Population genetics and evolution
    natural selection
    gene flow
    mutation
    drift or isolation
	2-3	ch. 2 (58-70)

✓ 4. Quantitative genetics	3	ch. 4 (117-127)

✓ 5. Variation in natural populations
    detection
    measurement
    magnitude
    examples
	4-5	ch. 3 (76-111)

✓ 6. Artificial selection
    types, including exotics
    procedures
    heritability estimation
    genetic gains
	5-6	ch. 4 (127-139)
		ch. 5 (143-162)

7. Selection and progeny testing	6	ch. 8 (229-264)
8. Breeding methods and results	7-8	ch. 8 (229-264)
9. Advanced generation breeding and hybridization	8	ch. 11 (345-367) ch. 13 (416-432)

10. Breeding for disease and insect resistance
    methods
    examples
    other special traits
	8	ch. 9 (270-298)

11. Mass production of seed and seedlings
    vegetative propagation
    seed production areas
    seed orchards
    inbreeding
	9	ch. 6 (167-208)
		ch. 10 (310-337)

12. Cone and seed collecting, processing, certification, records	9	ch. 6
13. Economics of tree improvement	10	ch. 14 (438-455)

FOR 410

14. Regional tree improvement program reviews   10   ch. 16 (480-491)

15. Special techniques in tree improvement   1-10

Hour exams are scheduled for: Wednesday 10/21
                                 Monday   11/16

Final exam: Friday, 12/11, 7:45-9:45 a.m.

Field trips scheduled: Monday, 10/12, Tree Research Center (Campus)
                          Monday, 10/19, Tree Research Center (Campus)

                          Saturday, 11/14, Kellogg Forest

                          Monday, 11/23, Tree Research Center (Campus)

FOR 410

Tree Improvement Laboratory Schedule

Period	Date	Topic
1	9/28	Overview of tree improvement.
2	10/5	Term project planning and lecture.
3	10/12	Nursery-greenhouse practices in tree improvement - Tree Research Center.
4	10/19	Measurement of genetic variation - Tree Research Center and lab.
5	10/26	Vegetative propagation: rooting, grafting, and tissue culture.
6	11/2	Tree reproduction and breeding techniques.
7	11/9	Term project planning and review.
8	11/14	Trip to Kellogg Forest genetic plantations. (No lab on 11/16)
9	11/23	Genetic record systems - seed processing. (Tree Research Center and laboratory).
10	11/30	Regional tree improvement program reviews.

FOR 410

FOREST TREE IMPROVEMENT

References

Will be placed on reserve in Room 218 or Room 36, Natural Resources Bldg. as needed.

Anonymous. 1974. Seeds of woody plants in the United States. U.S. Dept. of Agr., Forest Service. Agricultural Handbook No. 450. 883 p. (p. 41-166).

Wright, J.W. 1976. Introduction to forest genetics. Academic Press, N.Y. 463 p.

Dorman, K.W. 1976. The genetics and breeding of southern pines. U.S. Dept. Agr., Forest Service, Agr. Handbook No. 471.

Various authors and dates. Genetics of .... (each important tree species). U.S.D.A., Forest Service Res. Papers WO-1 to 35.

Proceedings of Canadian and U.S. Tree Improvement Conferences (e.g. Northeastern, North Central, Western, Southern, etc.)

FAO United Nations. 1978. Genetics. Unasylva 30:1-60. Second World Consultation on Forest Tree Breeding. Unasylva 24:1-132.

Annual Reports of the various regional tree improvement cooperatives.

Miksche, J.P. (Ed.) 1976. Modern methods in forest genetics. Springer-Verlag, Berlin. 228 p.

Hanover, J.W. 1980. Breeding forest trees resistant to insects. pp. 487-511. IN: Maxwell, I.G. and P.R. Jennings (Eds.) Breeding plants resistant to insects. John Wiley & Sons, N.Y.

Stern, K. and L. Roche. 1974. Genetics of forest ecosystems. Springer-Verlag, N.Y. 330 p.

Guries, R.P. and H.C. Kang (Eds.) 1980. Research needs in tree breeding. Proc. 15th North Amer. Quant. For. Genetics Group Workshop. 136 p.

Becker, W.A. 1975 (3rd Ed.) Manual of quantitative genetics. Washington State University, Students Book Corp., Pullman, WA 170 p.

Stanley, R.G. and H.F. Linskens. 1974. Pollen biology, biochemistry, management. Springer-Verlag, N.Y. 307 p.

Namkoong, G. 1979. Introduction to quantitative genetics in forestry. U.S. Dept. Agr., Forest Service Tech. Bull. No. 1588. 342 p.

FOR 410

Daniel, T.W., J.A. Helms, and F.S. Baker. 1979. Principles of silviculture. (Chap. 15, Tree Improvement). McGraw-Hill, N.Y. 500 p.

Todo, R. (Ed.) 1974. Forest tree breeding in the world. Government Forest Experiment Station of Japan, Meguro, Japan. 205 p.

Brown, A.G. and C.M. Palmberg (Eds.) 1978. Third world consultation on forest tree breeding. Vols. 1 and 2, Canberra, Australia. Commonwealth Government Printer, Cambera. 1235 p.

Gregorius, H.R. 1985. Population genetics in forestry. Lecture Notes in biomathematics, Vol. 60, Springer-Verlag, Berlin. 287 p.

Briggs, D. and S.M. Walters. 1969. Plant variation and evolution. World University Library, McGraw Hill, 256 p.

Gerhold, H.D., E.J. Schreiner, R.E. McDermott, and J.A. Winieski (Eds.) 1966. Breeding pest-resistant trees. Pergamon Press, N.Y. 50 p.

Bingham, R.T., R.J. Hoff, and G.I. McDonald (Eds.) 1972. Biology of rust resistance in forest trees. U.S.D.A., Forest Service Misc. Publ. No. 1221. 681 p.

Hanson, W.D. and H.F. Robinson (Eds.) 1963. Statistical genetics and plant breeding. Nat. Acad. Sci. Nat. Res. Council Pub. 982, Washington, D.C. 623 p.

Mettler, L.E. and T.G. Gregg. 1969. Population genetics and evolution. Prentice-Hall, Englewood Cliffs, Inc., N.J. 212 p.

Brewbaker, J.L. 1964. Agricultural genetics. Prentice-Hall, Inc., N.J. 156 p.

Allard, R.W. 1960. Principles of plant breeding. John Wiley & Sons, Inc., N.Y. 485 p.

Faulkner, R. (Ed.) 1975. Seed orchards. Forestry Commission Bull. No. 54, London. Her Majesty's Stationary Office. 149 p.

Hartl, D.L. 1980. Principles of population genetics. Sinauer Assoc., Inc. 488 p.

Owens, J.H. and M.D. Blake. 1985. Forest tree seed production. Canadian Forestry Service Information Report Pl-X-53. 161 p.

Willan, R.L. 1985. A guide to forest seed handling. FAO, United Nations, Rome, Italy. 379 p.

Anonymous. 1983. Reproduction of conifers. Can. For. Service Tech. Report 31. 38 p.

FOR 410

Other useful references:

Davidson, H. and R. Mecklenburg. 1981. Nursery Management. Prentice Hall, Inc. New Jersey. 450 p.

Williams, R.D. and S.H. Hanks. 1976. Hardwood nurseryman's guide. U.S.D.A., Forest Service Agr. Handbook No. 473, 78 p.

Douglass, B.S. 1969. Collecting forest seed cones in the Pacific Northwest. U.S.D.A., Forest Service Pacific N.W. Region. 16 p.

Furuta, T. 1968. Nursery management handbook. Univ. Calif., Agr. Extension.

Wang, B.S.P. 1974. Tree-seed storage. Can. For. Serv., Dept. of Environment Pub. No. 1335, 32 p.

Hedlin, A.F., H.O. Yates, D.C. Tovar, B.H. Ebel, T.W. Koerber, and E.P. Merkel. 1980. Cone and seed insects of North American Conifers. Canadian Forest Service, Ottawa. 122 p.

Franklin, E.C. 1981. Pollen management handbook. U.S. Dept. Agr. Handbook No. 587, 98 p.

Van Buijtenen, J.P. (Ed.) 1983. Progeny testing of forest trees. Proc. Workshop on progeny testing, June 15-16, 1982, Auburn, Alabama. South. Coop Series Bull. No. 275. 68 p.

Duryea, M.L. and T.D. Landis (Eds.). 1984. Forest nursery manual: production of bareroot seedlings. Martinus Nijhoff/Junk, The Hague. 385 p.

FOR 410

Publications Which Often Contain Original Research
On Forest Genetics

Silvae Genetica

Forest Science

Canadian Journal of Forest Research

Journal of Heredity

Hereditas

Physiologia Plantarum

Canadian Journal of Botany

American Journal of Botany

Botanical Gazette

Southern Journal of Applied Forestry

U.S.D.A., Forest Service Experiment Station Notes,
  Papers, and Technical Bulletins.

Proceedings of special symposia, regional conferences,
  and meetings

Tree Planters Notes

Unasylva (FAO, United Nations)

Theoretical and Populational Genetics

Forestry Chronicle

World Consultations on Forest Genetics (International Union of
  Forest Research Organizations)

Other useful sources of information: see list on previous pages.

FOR 410

Tree Improvement Regions

1. Inland Empire (N. Rocky Mountains)

2. West Coast: Washington - Oregon - British Columbia

3. California

4. Southwestern (S. Rocky Mountains)

5. Southeastern

6. Western Gulf

7. Northeastern

8. Lake States

9. Central States

Factors to consider in describing regional programs:

- Major species and their silviculture

- Land ownership

- Geneticists and their contributions

- History of forest genetics research

- Details of improvement procedures

- Progress

FOR 410

Questions to Address in Regional Tree Improvement Program Reviews

1. What species are being emphasized?

2. What organizations do tree improvement?

3. Who are some of the forest geneticists involved in Tree Improvement?

4. Are there cooperatives? How are they organized?

5. How is selection practiced?

6. Are there provenance tests? What species?

7. Are there progeny tests? What species?

8. What type of seed orchards are prevalent?

9. Is there an advanced generation breeding program?

10. Are the seed orchards progeny tested? What designs are used?

11. Acres of seed orchards? Seed production areas? Amount of improved seed produced?

12. Is there proof of genetic gain? What is the gain?

FOR 410

FIRST EXAM

NAME K.S. Bangarwa, Indian.

(20) 1. Define the following terms:

a. monoecious   Male and female flowers separate on the same plant.

b. tetraploid   Four basic set of chromosome. e.g. Birch   4n

c. cline :   Due to geographic / environmental clinal variation. Variation due to environment

d. haploid   Single set of genome (Basic ch) n

e. multiple alleles :   More than two alternative forms of allels at a particular locus. A B i for blood group in human being.

f. disruptive selection : Rejecting

By rejecting popn will be disruptive.

g. random drift :   Random change in popn & size. Random drift cause change in gene and genotypic frequency of the population

Page 1

h. population genetics: The science deals with quantitative traits with respective gene and genotypic frequencies of the community having common gene pool. Application of Hardy Weinberg law.

i. electrophoresis: It is to know structure and function of gene.

j. accelerated-optimal-growth
Increase in optimum growth due to selection

(6) 2. List three ways (i.e. vehicles, vectors, etc.) by which gene flow may occur in a tree species:

1. Wind
2. Insect
3. Migration

(12) 3. List 4 distinct goals or objectives that seem to be common to most tree improvement programs around the world:

1. High volume or high economic yield.
2. Pest resistance.
3. To introduce genetic variation if not in wild
4. Creat dependable sources of seed production

(4) 4. What type or system of silviculture is required in order to utilize genetic methods for improving forests?

Better management which includes soil, fertilization in order to get high quality wood in a short time and inexpensive in presence of better breeding line.

Page 2

(4) 5. The predominant chromosome number for species of pines is 2n = __24__.

(15) 6. With rare exceptions (e.g., Red pine) inbreeding is detrimental to the normal growth and development of trees.

   a. What is the genetic reason or mechanism why this statement is true?

   Homozygous lethal effect. ie lethal effect is not express in heterozygous condition

   b. If a tree self fertilizes, what would be the F value or inbreeding coefficient? __1__

   c. Repeated generations of even mild inbreeding in a population of trees will lead to increased __homozygosity__ and decreased __heterozygosity__ at all gene loci.

   d. How do tree breeders attempt to minimize the possibility of inbreeding in a long term tree improvement program?

   −1 *Random mating*
   Avoid selfing

(10) 7. For each of the following genetic conditions or mechanisms indicate whether it tends to increase (+) or decrease (−) genetic variability by its presence or action:

   (+) a. polyploidy
   (−) b. linkage
   sav ( ) c. mitosis   *No change in normal mitosis.*
   (+) d. meiosis
   −1 ( ) e. regulating genes
   (+) f. multiple alleles
   ( ) g. mitosis   *No change*
   (+) h. independent assortment
   (+) i. segregation
   (−) j. dominance

(4) 8. There are at least four reasons why almost all of the commercial bareroot tree nurseries of Michigan are located on the west side of the state. Give 2 of these reasons.

   ✓ 1. *Geographic.*

   2. *Flowering time.*

(10) 9. In an analysis of recombination and crossing over involving 4 genes, R = resistant, F = leaf color, S = straight stem, and C = 3-carene, the following recombination values were measured:

Between R and S: 32%
F and C: 18%
C and R: 29%
S and C: 3%

R ←—— 32 ——→ S
←—— 29 ——→ C ←→ 18 ——→ F
        3

a. Show the arrangement of these 4 genes on the chromosome.

R ←—— 29 ——→ C S ←—— 18 ——→ F
←—— 32 ——→   ←→
                 3

b. What is the % recombination between F and R? __47%__

c. Gene mapping is based on what three principles of genetics?

1. % cross over value in test cross.

2. Linked gene or Linkage or chromosome on same chromosome

3. Sequence of gene on chromosome.

(3) 10. A. What is the main reason for doing a provenance test as a first step in a tree improvement program?

To know the adaptability in new environment. If better adapted with desireable traits then time required will be comparatively very less for improvement.

(12)   B. List four of the many basic steps involved in a complete provenance test for a tree species.

1. Collection of seed: Collect the seed before shedding send for testing with complete information. like location Height etc.

2. Grow the tree species in different areas for adaptation and evaluation.

3. Record the various traits for evaluation.

4. Summarize the performance by statistical analysis of various traits.

Page 4

Collection, maintenance, evaluation and distribution

FOR. 410

Second Exam

1. Give the formulas for calculating (predicting) genetic gain from selection. Explain in detail how each factor in the formula is derived and what each tells the tree breeder about procedures to follow for improving a species.

2. Assume that you are a forest geneticist beginning the first work ever done towards improving the state tree of Michigan, eastern white pine. Provide a step by step description of what you would do to, (1) determine the pattern and amount of genetic variation in the species, (2) determine the sources of genetic variation, and (3) determine the amount of genotype x environment interaction.

3. What are the genetic advantages and disadvantages of clonal tree improvement and reforestation practices for a species that is amenable to vegetative propagation?

4. Explain why it is important to establish a "broad genetic base" in a tree improvement program and tell how this is accomplished.

K. S. Bangarwa

Question No 1

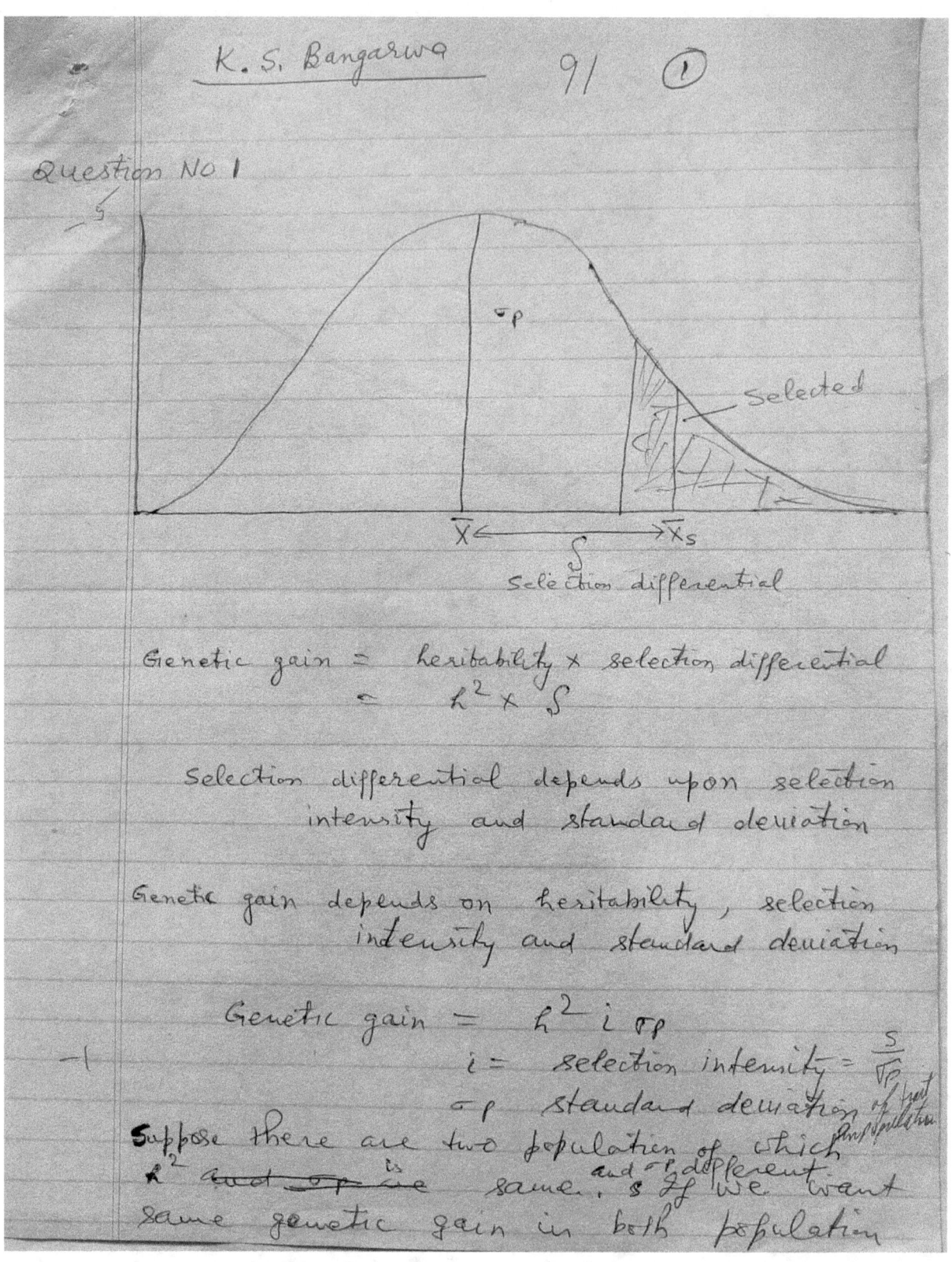

Selection differential

Genetic gain = heritability × selection differential
= $h^2 \times S$

Selection differential depends upon selection intensity and standard deviation

Genetic gain depends on heritability, selection intensity and standard deviation

Genetic gain = $h^2 \, i \, \sigma_P$

i = selection intensity = $\frac{S}{\sigma_P}$

σ_P = standard deviation of that population

Suppose there are two population of which h^2 is same and σ_P different. If we want same genetic gain in both population

(2)

we have to select more plants where σp is less.

Freq

Ht

Selection should be specific.

If we want same selection differential having different σp

More σp less i (selection intensity)

Less σp more i. Yes

Define h^2
How is h^2 derived?

Q No 2

(i) Determine the pattern and amount of genetic variation — Effect of North and South

(1) I will first decide objective of eastern white pine breeding.

(2) Collect the seed [how much/bulk?] from different sources depending upon objectives of breeding.

(3) Grow the different seed lots in different regions (sites), ~~different stands and different family~~.

(4) Collect the data from all over planting and evaluate.

(5) The variation will be (i) Between regions ~~Four types~~ This is a provenance/ (ii) Between stands ~~Little variation~~ progeny test OK (iii) Between families

Select the best region.
Best stand in best region.
Best family in best stand.

ANOVA ~~variance~~

S.V.	Hf
Between region	X
Bet Stand	Y
Bet family	Z

Variance ~~reflect~~ reflects amount of variation.

N

Plants

S

After study we will know the pattern of variation about growth etc and

After knowing the pattern of variation and amount of variation we can goes for to further breeding accordingly.

Determine sources of variation:
There are five types of variation in trees
(1) Geographic (Provenance) variation
(2) Site variation = environment or test location?
(3) Stand variation
(4) Individual or family variation
(5) with in individual variation

The variation can be worked out by growing in different region site and stands.

ANOVA	variation
Sources of variation	Mean Sum of Square
Between Provenance. About 30%	It is more in adaptable character than growth
Between site.	very less
Between stands.	very less
Between individuals. about 70%	It is more in growth character than adaptable.
with in individuals.	very less.

Variation is the best tool of selection

(3) Determine the amount of genotype × environment interaction.

Grow the different a collection in different environment with RBD design.

| | E_1 R_1 R_2 R_3 | E_2 R_1 R_2 R_3 | E_3 R_1 R_2 R_3 | E_4 R_1 R_2 R_3 |

Collection I (genotype)
Collection 2
...
collect n

ANOVA for genotype × environment interaction

Source of variation	Variation
Replication	x
Collections (Genotype)	y
Environment	z
Genotype × environment	a

Genotype × environment interaction: interaction of a particular genotype in a particular environment.

QNO 3   Advantages:- (1) We can achieve achieve the improvement very fast if improved clone is available.

(2) We can utilize the specific combining ability which is otherwise difficult to utilize. The best performing F1 can reforested by vegetatively.

(3) Vegetative require less time for establishment.

(4) Spruce produce late flowering that is difficult other than vegetative propagation

(5) It is easy to plant by grafting where seed development is poor.

-1   Uniformity

Disadvantage (1) We can utilize only existing variability of natural stands for species improvement.

(2) Clonal tree improvement will lead narrow genetic base ie harmful if there is infection of disease than whole if will destroy

3. We cannot add new character to the population by clonal free improvement.
       outbreed
         inbreeding

Q No 4 — **Importance of broad genetic base**

Broad genetic base means having maximum genetic variability. It is effected by mutation, gene flow, selection and random drift. Selection is the major force to disturb broad genetic base.

Importance: Suppose we are continuously selecting a population which leads to narrow genetic base although we are selecting towards our objectives. Suppose we achieve our target. If suddenly there appears any disease in your strain, then you will not be able incorporate resistance with out broad genetic base. No further improvement will be possible. Broad genetic base will certainly provide some trees suitable to your objective ~~also avoids inbreeding~~

1. Broad genetic base can be accomplished by diverse mating, not at first selecting heterozygous individual
2. Recurrent selection (select and intermate repeated) ~~will be~~ is helpful in maintaining broad genetic base

Avoid selfing = inbreeding
No continuous clonal tree improvement

NAME __K.S.B.__

FINAL EXAM

(30) 1. Below are some facts about three tree species which you are to use in designing a tree improvement plan through to production of F_3 seed. In other words, show the procedures you would recommend using a clearly annotated flow diagram for <u>each</u> of the species (3 plans). Be sure to indicate when improved seed will be produced for commercial use.

Species	Age to flowering	Trait to improve	h^2 of trait	σ_p of trait	Ease of vegetative reproduction	Rotation cycle (use)	Major Source of σ^2_G
A. Black Walnut	20 yrs.	height-diameter (veneer)	8%	high	Grafting only	40 yrs.	Regions and Families
B. Black Locust	3 yrs.	stem straightness (lumber)	30%	high	Relatively easy by cuttings	25 yrs.	Individual Parent trees
C. Red Pine	10 yrs.	volume	5%	low	Grafting only	40 yrs.	Individual Parent trees

(10) 2. Describe how you would go about locating and developing a seed production area for any given species.

(15) 3. Assume that you are a tree improvement specialist for a large timber corporation. Your first assignment is to establish a new seed orchard by grafting using 15 clones. Develop a step-by-step plan for **(1)** producing the planting stock so as to be ready at the right time, **(2)** selecting the proper site for the orchard, **(3)** arranging ramets in the orchard, **(4)** managing the orchard for maximum seed production, and **(5)** harvesting the seed.

(10) 4. Define and distinguish between the following methods of selection; especially in terms of their consequences:

    A. Mass selection
    B. Family selection

(10) 5. Briefly tell each of the processing steps involved in producing pine seed from cones collected on the ground.

(15) 6. Select four different types of mating designs that are commonly used in tree breeding and show how each is constructed (diagram). For each design give one advantage and one disadvantage over others you have chosen.

(10) 7. Explain the difference between Mendelian inheritance and quantitative inheritance for the benefit of a lay person (e.g., the public).

3 Answer Sheets of Tree Physiology Course Examination with B Grade in USA

73½

Name Kulvir Singh Bangarwa
Faculty of India

FORESTRY 411

FIRST EXAM

(2) 1. Show by a simple diagram of a portion of a tree stem cross section how the xylem cell pattern would appear if cambial cells divided only periclinally. Because this never occurs in nature, your diagram is, of course, strictly hypothetical.

−2

[diagram: rectangle → rectangle with dashes → rectangle with hatching labeled "xylem"]

Fusiform initial

(2) 2. When do cambial cells divide in the anticlinal plane?

−2   Normal growing period (starting slow)

[diagram: rectangle → rectangle divided → small rectangle]

(2) 3. How do we know that it is usually only the single cell circle of cambial initial cells around the stem that divide anticlinally?

opposite division
others have [diagram of rectangles with hatching]

(4) 4. The blue haze often seen over forested regions on sunny days results from atmospheric reactions involving __waxy terpenes__ compounds released by trees. These compounds are produced in specialized cells called __Cuticle epithelial__ cells in different tissues of a tree.

−3

(4) 5. Two possible functions for epicuticular waxes on tree leaves are:

1) ~~wax~~ Avoid loss of water and
2) Protection . In conifers and many hardwood species epicuticular waxes are most closely associated with blue color tree .

(6) 6. What are the two types of reasoning used in the scientific method?

1) __Inductive__ __General law__

2) __Deductive__ __law to new problem__

Which of these is the "safest" or most conservative in terms of not making an error in formulating a hypothesis? __Inductive__

−1 Which of these is most likely to result in major advances towards the solution of major scientific problems? __Deductive__

What is meant by the term "strong inference?" __Presence or absence of one event cause major deviation to other.__

(10) 7. Give 5 differences between roots and shoots of a tree in terms of their developmental anatomy and morphology.

	Root	Shoot
1)	Root cap is there	No cap.
2)	No bud and scale	Bud & scales are there
3)	No node.	Node is there.
4)	No pith	Pith is there.
5)	Root hair	No root hairs.

(2) 8. What is the reason that the bark of a tree is easily removed after growth begins in spring or early summer whereas it is very "tight" during fall and winter?

Fresh growth is soft so less intact

Soft bark. so easy to remove in spring

(6) 9. Briefly describe three kinds of experimental evidence which support the auxin theory for initiation of cambial division in a tree.

1) Auxin is produced in spring by leaf so cambial division starts after spring. and activity stops in october.

2) Auxin is more in crown area a auxin gradient from top to bottom. Cambial activities more in crown and less in clear bole.

−1 3) Chilling effect is required to produced auxin? If temp rise in Feb–March then there will be growth not in Nov.

(10) 10. In contrast to a diffuse porous tree species a ring porous tree will:

   a) have a (wider, na<u>rr</u>ower) sapwood radial thickness. ✓

   b) have a (<u>more</u>, less) rapid rate of water movement up the stem.

−2 c) have (more, <u>less</u>) annual rings in the functional sapwood.

   d) begin cambial activity (<u>earlier</u>, later) with respect to bud activity. ✓

   e) be (<u>more</u>, less) subject to physiological damage due to stem injury.

(4) 11. Name two ring porous tree species whose survival is being threatened by diseases or insects that primarily affect their water conducting system.

−4 1) __Spruce ?__
    2) __D. fir ?__

(2) 12. The cambium of northern tree species (does, <u>does not</u>) ✓ require a chilling treatment for resumption of growth after favorable conditions are imposed.

(2) 13. Auxin and gibberellic acid exert their regulatory affects on cambial cell...

    a) division.
    b) enlargement.
    c) secondary wall formation.
    d) all of these ✓

(2) 14. In partitioning photosynthate within a tree stem wood (xylem) has a ___ (high, low) priority relative to that of roots and leaves.

(14) 15. Figures 1, 2 and 3 on the following pages show three anatomical systems in a tree. For each number on a figure, define the tissue at the tip of the arrow according to standard terminology.

    Fig. 1
      1) Apical meristem A I
      2) Mother cell (cotyledons)
      3) Mother cell
      4) Procambium
      5) elongation zone
      6) Rt
      7) Maturation zone

    Fig. 2
      1) Apical Meristematic
      2) Parenchyma
      3) wall

    Fig. 3
      1) ~~Hypocotyl~~ Apical Needles of conifer
      2) Stem of conifers epicotyl
      3) Primary leaves
      4) Hypocotyl

(8) 16. In contrast to the main stem of a tree, branches usually die as the crown moves upward. List 4 physiological events that contribute to the senescence and death of lower branches.

    1) Lower branch have less foliage

    2) Ring ~~not able to reach~~? terminate before reaching

    3) Less sunlight

    4) Less photosynthesis

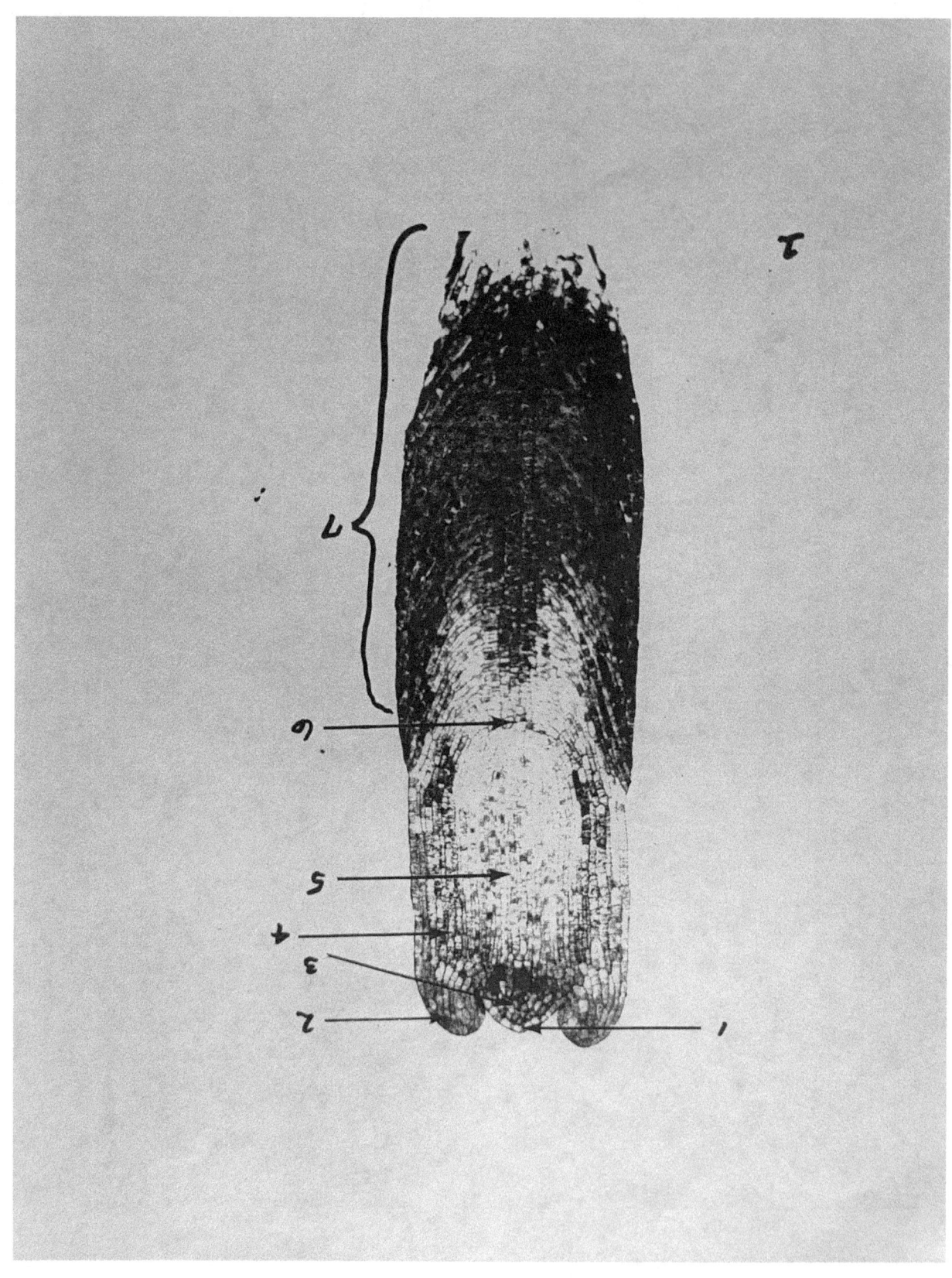

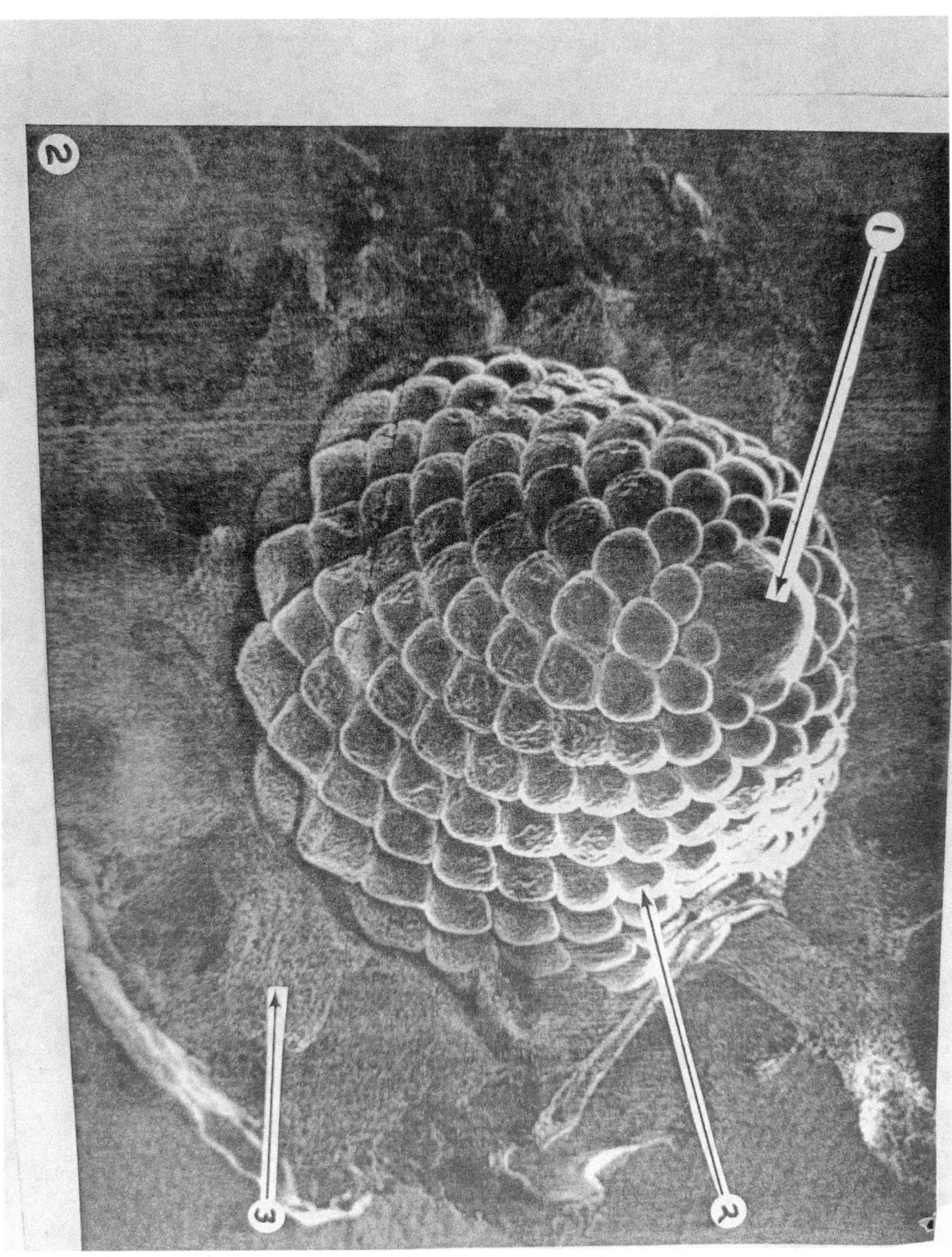

(2) 17. The wood of rapidly growing trees (wide rings) is likely to be (higher, lower, no different) in specific gravity than that of slower growing (narrow rings) trees.

(10) 18a. Show the correlative relationships between the following factors: leaf area, sapwood basal area, basal area growth for the last 5 years, and total stem wood volume growth.

$$\frac{\text{Basal area growth for the last 5 yrs}}{\text{Sapwood basal area}} \quad \frac{\text{Total stem wood vol growth}}{\text{Leaf area}}$$

18b. How would the relationship between total leaf area and sapwood basal area be expected to vary between trees growing on droughty versus moist sites?

Leaf area will be less in drought condition

18c. Why? *More leaf foliage require more water*

−1

D.fir $\frac{LA}{SA}$ = .14 Drought
.75 moist
hit = .34
const = .59

(4) 19. Give two reasons why the ray cells of a tree stem are important to the vigor of the tree.

1) *Ray cell responsible for sugar.*

−1 2) *Basal growth will be more in presence of ray cell.*

(4) 20. List 4 ways in which the heartwood of a tree differs from the sapwood.

−1 1) *Heartwood have no living function.*
   2) *Cell act as dead.*
   3) *Ray cell decreases.*
   4) *Sugar decreases.*

85

Name Kulvir Singh Bangarwa

FORESTRY 411
Second Exam

(103 total points)

(75) 1. For any given north temperate tree species indicate by <u>checking</u> the appropriate column whether a specified change in plant or environmental characteristic is <u>most likely</u> to increase, decrease or have no affect on <u>leaf net photosynthetic rate</u>, all other factors are held constant.

−15

Change in plant or environment	Net photosynthetic rate in mg CO_2 / cm^2 / hr		
	Increase	Decrease	No change
1. Increase leaf area	✓		✓
2. Increase stomatal density	✓		
3. Increase cuticle thickness		✓	
4. Decrease stomatal waxes	✓		
5. Change from no wind to gentle wind	✓		
6. Decrease stomatal resistance	✓		
7. Change from leaf plastochron 1 to 4	✓		
8. Change from leaf plastochron 4 to 30		✓	
9. Increase mesophyll resistance		✓	
10. Increase from 15°C to 25°C	✓	✓	
11. Decrease relative humidity from 50% to 20%		✓	
12. Increase RUDP carboxylase	✓		
13. Increase light from 8,000 f.c. to 15,000 f.c.	✓	✓	
14. Change from sun leaf to shade leaf	✓	✓	
15. Decrease photorespiration	✓		
16. Increase mitochrondrial respiration		✓	✓
17. Increase in length of growing season	✓		✓
18. Change O_2 level from 21% to 10%	✓		
19. Change CO_2 level from 500 ppm to 2000 ppm	✓		
20. Increase disease on leaf		✓	
21. Increase disease on root		✓	✓
22. Increase tree age 30 years		✓	
23. Change light from 450 (blue) to 600 nanometers (green)	✓	✓	
24. Increase light compensation point	✓		
25. Increase CO_2 compensation point		✓	
26. Change from C_3 to C_4 characteristics	✓		
27. Decrease in phloem translocation rate		✓	
28. Increase transpiration rate without drought	✓		
29. Sunken versus surface stomates		✓	
30. Leaf display more vertical		✓	
31. Increase in leaf intercellular spaces	✓		
32. Increase in fruit production		✓	
33. Decrease in leaf hairs	✓		

Page 2
411 Exam

Change in plant or environment	Net photosynthetic rate in mg CO_2 / cm^2 / hr		
	Increase	Decrease	No change
34. Increase in xylem translocation rate	✓		
35. Lower crown to upper crown branch	✓		
36. Increase ratio of leaf area to sapwood area	✓		
37. Decrease chlorine		✓	
38. Increase manganese	✓		
39. Move shade tolerant species to higher light	✓		
40. Move shade intolerant species to lower light		✓	
41. Leak in the IRGA of a closed system		✓	
42. Increase number of chloroplasts	✓		
43. Increase in ratio of ADP to ATP	✓		
44. Remove temporary standing water around tree base	✓ considered flood	✓	✓
45. Decrease PAR		✓	
46. Increase epicuticular waxes		✓	
47. Transplant seedling		✓	
48. Gently shake seedling briefly each day	✓	✓	
49. Apply antitranspirant		✓	
50. Apply abscisic acid	✓	✓	

(8) 2. List four important factors that contribute to the control of branch angle of a tree and tell how each acts.

a) Bud insertion: → Bud first grow up side then come down.
Branch maintain angle with stem, & secondary branch with primary branch.

b) <u>Plagiotropism</u>: It is the position of grafted that maintain characteristics. The form depends on

Page 3
411 Exam

c) Preconditioning: Compression wood depends on preconditioning if branch is bending down than CW will be lower side and if bending up than upper side. CW will be on opposite if CW is already exist.

d) Gravitational stimuli: It is the geotropic effect. Shoots has negative geotropism.
Clinostat effect is best example of gravitational stimuli.

e. Equilibrium: Tree maintain equilibrium if branch change from equilibrium then Reaction wood formed to maintain equilibrium.

(20) 3. Define each of the following terms clearly.

a. plagiotropism: This refers to the position of grafted. The branch angle

b. epinasty: It referes to control of apical to lateral branches.

c. tension wood: in reaction wood — It formed in angiosperm. It formed on upper side of leaning stem. It pulls the branch. It may be due eccentric cambial growth.

d. clinostat: It referes to change in position with respect to gravitational force. Compression wood change if there is clinostat.

e. apical control: It is the control of terminal bud over lateral branches. If no apical control then plant will be bussy.

f. apical dominance: → It is most common in conifers. The tree form excurrent. Terminal stem is always leading lateral branches.

g. seismomorphism: It refers to mechanical response. eg. Tree has effect of wind or shaking. Shaking cause increase in stiffness. diameter increase height decrease.

h. topophysis: It relates to plagiotropism. Grafted branch continue its characteristics after grafting on tree. Topophysis also relates to flower production.

i. gravimorphism: → morphology because of gravitational force. New growing buds show negative geotropism. Compression wood will be on upper side if grown up down

j. cyclophysis: refers to maturation of meristems on tree. It also relates to topophysis. It is internal effect